Into a Black Hole

Safiyah J Dandashi

Copyright © 2013 Safiyah J Dandashi

All rights reserved.

ISBN: 1492207764
ISBN-13: 978-1492207764

DEDICATION

To my dad for taking us to the *Almost Heaven Star Party* in the darkest place on the Eastern Coast of the United States of America.

TABLE OF CONTENTS

Chapter I 1

Chapter II 3

Chapter III 7

Chapter IV 12

Chapter V 14

Chapter VI 16

Chapter VII 18

Safiyah J Dandashi

ACKNOWLEDGMENTS

I'd like to give special thanks to Mrs. Eva, my writing teacher, and to *Microsoft Word's Clip Art* for some of the pictures in my book.

CHAPTER I

In the year 3333, Daniel was watching TV when he spotted a commercial for *Wheel of Fortune,* advertizing a special week for children. Kids will compete in groups of four, and that special week will be called the *Group Kids' Week*. Daniel immediately called his friends Annie, Samuel, and Jennet.

"Do you want to take the quiz with me for *Wheel of Fortune?*" Daniel asked his friends.
"Sure!" they answered. A few minutes later, the doorbell rang.
"I'll open it!" yelled Daniel as he ran down the stairs.

It was Annie, Samuel, and Jennet. They went up to Daniel's room and he turned on his computer. He went to the website shown in the commercial and clicked the button to start the quiz.

....

After completing the quiz, printed on the screen were these words, "PLEASE WAIT FOR RESULTS"

CHAPTER II

"CONGRATULATIONS! YOU PASSED!" was printed on the screen next to a picture of the host standing beside the wheel. The four friends jumped up and down screaming, "WE PASSED! WE PASSED!" Daniel ran to his mother and told her the good news.

"When do you play?" she asked, taking out her computer to buy tickets.
"We play in a week. It comes with ten free tickets that should arrive tomorrow!"
"Where is it?"
"It's here, in the Kennedy Center!"
"Great!"

On the night before the game, the four friends couldn't sleep because they were so excited. In school, Daniel couldn't pay attention to the teacher because he couldn't stop thinking of the night's game.

Later in the day, at Daniel's house, a few minutes before they left for the game, Annie, Samuel, and Jennet came over, and Daniel's mom offered to drive them to the game. They accepted the offer and piled into her car. They had been studying all week and couldn't wait to play. They drove to the Kennedy Center without incident and went to the room directed………

"Final Spin!" exclaimed the host. He spun the wheel………

In the final commercial, the friends were excited. "We have the greatest amount of money. I think that means we walk up to the small wheel and solve a puzzle," whispered Samuel………

Their family members ran out on the stage as the friends won a trip to another planet. The trip was to take place in a week from the day they won.

………………………………………………

In a few days, they were ready to leave Earth, but they had to wait until the next day. They were all restless during the night. In the morning, when his mom woke him and his friends up, Daniel shot out of bed, gobbled up his breakfast, and changed into his everyday clothes before one could blink. Today was the day into outer space!

"Daniel, slow down!" shouted his mother as his

friends yawned and slid out of their beds. To pass the time while his friends got ready, Daniel took out his flexible, walking iPod and started to play *Temple Run*. When his friends had finally finished breakfast, they walked to the space shuttle and boarded it. Daniel sat in the commander's seat. He had trained for it throughout the week.

Now, he was ready. In a few minutes, a robotic voice exclaimed, "T-MINUS SIXTY SECONDS." In fifty seconds, they heard the voice again, "T-MINUS TEN, NINE, EIGHT, SEVEN, SIX, FIVE, FOUR, THREE, TWO, ONE, BLAST OFF!"

An incredible noise filled the cockpit! The four friends were pressed into their seats! Then suddenly, it was over. They were in outer space. They looked back and saw Earth disappearing behind them. They whizzed past the moon and Mars *and* through the asteroid belt in a few minutes.

In a day, they were nearing the center of their galaxy, the Milky Way. Suddenly, these words appeared on the monitor:

YOU ARE NEARING A BLACK HOLE.

CHAPTER III

The four panicked. They didn't know what to do! Then, in an instant, they disappeared from this universe. They had gone into the black hole. It looked like a long black tunnel to them. In a few minutes, they emerged from the tunnel with a loud POP and crashed on the surface of a planet.

As they looked back at the black hole, it vanished. They took a sample of the air and discovered that it was identical to Earth's atmosphere. They decided to explore the planet. They stepped carefully out of the spaceship. They saw star shaped bushes with star shaped flowers. They saw star shaped trees with star shaped nests and star shaped birds. They went into a supermarket and saw star shaped food. "Wow," Annie whispered.

Suddenly, every star-person looked at the four friends. Then, they shifted shapes. They used to be stars, and now they looked like the four friends!

Then one star-man stepped forward. "Hello," he proclaimed. "From all the star-people living on this planet, only another star-man and I can speak your language. The other star-persons speak a different, alien language called *Ooombaien*. Our planet's name is *Ackattuoieaou*."

"Why do you speak our language?" asked Daniel.

"I speak your language because when people see a *UFO,* it is my spaceship looking for a field to land in."

"I then shape myself like children to not stand out as an 'alien'. I go to school, learn English, and come back so I can know your language in case your people come through the 'Disappearing Tunnel'. We can go to my house and I will call the mayor."

They followed the alien to one of the smaller star-buildings. Once inside, a female star gestured them to a table with four chairs and a box of cookies, all of which were shaped like stars. They sat down and helped themselves to the cookies. The star-woman soon came back with some cups and what appeared to be apple juice. She left the room and the star-man came back and exclaimed, "My name is Mr. Oombaro. The mayor will see you in half an hour." "What are your names?" questioned Mr. Oombaro.

"My name is Daniel," proclaimed Daniel.
"My name is Jennet," answered Jennet.
"My name is Samuel," exclaimed Samuel.
"My name is Annie," whispered Annie.

They all asked, "Do you know why the black hole disappeared?"

"The black holes on *this* side appear and disappear every six months. You came through as it was disappearing," answered Mr. Oombaro.

"You mean that we have to fix the spaceship in six months! We'd better start," exclaimed Daniel.

"When the mayor leaves, I'll call the spaceship builder to come help you," promised Mr. Oombaro.

Just then, the doorbell rang. "I think that's the mayor," Mr. Oombaro exclaimed. He opened the door.

"Oooiuyachaiuoe yacooooiueacha," stated Mr. Oombaro.

"Aiecoocha," answered the mayor, "Atchkl cchokla nonku?"

"Achuka," exclaimed Mr. Oombaro.

He turned to the children, "Kids, this is the mayor. I'll be your translator."

An hour later, the interview was coming to an end. The last question was asked, "Aka u chica?"
"When are you leaving?" translated Mr. Oombaro.
"As soon as we can," answered Daniel. Mr. Oombaro translated to the mayor, who nodded, stood, and exited. Mr. Oombaro followed to fetch the spaceship builder.

CHAPTER IV

The other star-man that could speak English on *Ackattuoieaou* was the spaceship builder. He was also the only one whose name was English. They found this out because the first thing he exclaimed to them was, "Hi! My name is Henry."

They immediately started working. Henry found out how to make the spaceship travel as fast as light. He also added heat-resistant plates that could let them go through a star! In addition, he put insulating foam that lets one survive in a place that is as cold as negative 150,115,060,159 degrees Fahrenheit, which would help if they happen to crash-land on a planet that is four light years away from its star, causing it to be as cold as an ice-cube in a freezer stuck in the ice age! But he could not fix the engine. Their hopes for returning evaporated.

Then, Henry's son, Jumka, shouted, "Jumkalunta tolamato tengune genpaolu!!!!!!!!!"

Henry translated, "I know how to fix the engine!!!!!!!!!"

"How?!?!?!?!?!" the kids inquired. As an answer, Henry's son took out some wires and fuses from the engine and replaced them. The engine started humming and continued to hum until Daniel jumped in the spaceship and turned it off.

"Yippee!!!!!!!!!!!!!!!!!!!!!!!!!!!!!!!!"Daniel shouted, "Now all we have to do is wait three months for the black hole to appear."

CHAPTER V

"Oh, no!" yelled Daniel's mother, "Look at this!" She was reading a newspaper which had a news headline that stated, "Space Shuttle Sucked into Black Hole!!!!!!!!!!!!!!!" She quickly read the rest of the article. Her eyes were glued to this sentence, "The missing are Daniel Smith, Annie Bahjaji, Samuel Neil, and Jennet Gauss."

"Aren't those Daniel's friends' names?"
"I think so," grumbled Daniel's father.

....

Three months later, Daniel saw some twirling in the sky that was turning black. "I think the black hole is appearing! Grab the spaceship!!!" he yelled. They all, including the star-people, ran out, and sure enough, there was the black hole.

Into a Black Hole

They scrambled into the spaceship and started the countdown:

"T-MINUS TEN, NINE, EIGHT, SEVEN, SIX, FIVE, FOUR, THREE, TWO, ONE, BLAST OFF!" shouted the aliens as the spaceship rumbled, shook, and blasted into the black hole. Darkness surrounded them as they flew in.

Meanwhile, on Earth, Daniel's mom turned on the television for the evening news. She gasped when she saw the screen. "Breaking News! NASA has spotted an alien spaceship emerging from the black hole!"

CHAPTER VI

"Do you think it's really them?" asked Daniel's mom as his parents drove to NASA headquarters.

"Could be," mumbled her husband sleepily.

....

Daniel spotted the Hubble Space Telescope and the International Space Station (ISS). He waved at the ISS and glimpsed someone waving back. Then, they entered Earth's atmosphere.

....

On the video in NASA's headquarters, Daniel's parents saw a spec come into the atmosphere. The spec grew larger and larger until they could see the face of the driver. It was Daniel! The shuttle landed BOOM,

Boom, boom. As soon as the four friends exited, their families mobbed them with hugs and kisses. After six months of worry, the four families were reunited.

CHAPTER VII

Eight months later, Daniel saw an ad for the *Group Kids' Week*, but this time it was on *Jeopardy*. "Here we go again!" he exclaimed as he picked up the phone to call his three friends...

Safiyah J Dandashi

ABOUT THE AUTHOR

Safiyah is a nine year old girl who is highly gifted in math. She enjoys writing but not *reading* her papers in front of her writing class. She loves astronomy, and reads all the books about it that she can get her hands on. Safiyah lives in Northern Virginia with her mother, father, and older sister.

Also written by the author are *The Mystery of the Planet Mix-Up* (2009), *The Mystery of the Disappearing Monkey* (2010), *Musings of a Seven Year Old* (2011), *Musings of an Eight Year Old* (2012), and *Musings of a Nine Year Old* (2013).

www.ingramcontent.com/pod-product-compliance
Lightning Source LLC
Chambersburg PA
CBHW042348200526
45159CB00034BA/849